Opioids Abuse

Dr. Altti Heikkinen

Kelsiehanson Verlag

DEDICATION

To My Wonderful Readers

CONTENTS

ACKNOWLEDGMENTS

Thank you to everyone who strives to grow and help others grow

1 ABSTRACT

The incidence of opioid abuse is an alarming issue within the healthcare industry as it could cause death and could increase the healthcare cost. Monitoring of opioid intake is essential and interoperable data is highly required to reach the goal of monitoring opioid intake using the Health Information System (HIS). Therefore, this study aims to explore the current workflow of HIS to curb the incidence of opioid abused in Saudi Arabia and to determine healthcare practitioner's perception towards the role of thecurrent HIS in mitigating the opioid abuse issues. Methods: A Case Study Research was performed at Emergency Department, King Faisal Specialist Hospital and Research Centre (KFSHRC) and Al-Amal Mental Health Complex (AAMHC) from 1st March to 20th April 2018 whereby an observation on the Health Information System (HIS) workflow process was conducted and 37 healthcare practitioners were recruited to determine their perception on the current HIS through 5 Liker-scale questionnaire. Results: Opioid abuse issue is a prevalent issue in Saudi Arabia. Although the healthcare practitioners perceived the current HIS positively in curbing the opioid abuse or misuse issue but only within their organization, the absence of interoperability in the current HIS could be observed. Hence the ultimate role of HIS to curb the opioid abuse issue might not be reached. Conclusion: In order to use the HIS in curbing the issue of opioid abuse, the current workflow of HIS in Saudi Arabia needs to be improved and the function of real-time health information exchange needs to be developed.

Index Terms— Opioid abuse, Misuse opioid, HIS, PDMP, Interoperability, Health Information Exchange.

2 INTRODUCTION

Recently, an increment of opioid prescriptions specifically for acute and chronic illness has been observed. This increment has subsequently caused the increment of the incident of opioid abuse. The United States Drug Enforcement Agency Records revealed that between 1997 and 2006, the sale of opioids was increased by 127% (Worley, 2012). Without proper monitoring of the prescription of opioid's drugs, it could relate to far-reaching consequences within the healthcare industry. For instance, as noted by Brat et al. (2018) several countries, including the United States and Saudi Arabia, have reported a significant increase in the number of deaths associated with abuse and misuse of prescription drugs, especially opioids. To illustrate this further, 75.2% of pharmaceutical overdose are reported to be related to opioid and almost 15,000 of United States citizen die due to the prescription painkiller overdose, yearly and recent data in 2018 highlights that 40% of all deaths associated with opioid overdose is caused by the prescription opioids(Jones, Mack and Paulozzi, 2006; Centers for Disease Control, 2011; Hadland et al., 2019).

The abuse of opioids could adversely affect patients' health by causing opioid use disorders (Denisco et al., 2011). Disorders caused by opioid abuse are almost quadruple times more common compared to disorders associated with other substances (Alblooshi et al. 2015). In addition, the incidence of opioid abuse also relates to a higher health care cost due to high rates of a physician visit, mental health inpatient and outpatient services and admissions. Also, the high cost is associated with increased rates of emergency department visits, motor accidents, trauma and substance abuse treatment. Past study has shown the cost for persons abusing opioid is 8 times more than the cost for non-abusers ($15,884 vs. $1830) (Hahn, 2011). In another study on Economic Burden of Opioid Abuse conducted in United States, it was highlighted that the opioid abuser needs to pay additional USD 10,000 to USD 20,000 for their healthcare cost yearly compared to non-abuser (Kirson et al., 2017)

Opioid abuse is commonly associated with the incidence of doctor shopping. The issue of doctor shopping is a prevalent issue within the healthcare industry specifically involving the prescription of opioid drugs (Worley, 2012; Patrick et al., 2016; Buchmueller and Carey, 2018). The issue of doctor shopping is specifically critical within the country with unintegrated Health Information System (HIS) whereby a patient has a higher tendency to manipulate their drug prescription's history and schedule as the accessibility of patient's medical history by a physician is limited (Simeone, 2017; Biernikiewicz, Taieb and Toumi, 2019; Lehne et al., 2019; Kruse et al., 2020).Although a country might develop a high advanced HIS, without having an interoperability system, the full capacity of HIS in curbing the opioid abuse might not be reached (Gandhi et al., 2016).In curbing the opioid abuse, the ability of the HIS to provide interorganizational information exchange is crucial as without the interoperability system, patient's medical history could be manipulated (Lupton, 2014; Kohli & Tan, 2016; Kellermann & Jones, 2013; Keller, Kelling, Cornelius, Oni, & Bright, 2015).

In a setting whereby there is a limitation on the interoperability of HIS data, missing a real times information about a patient medication's history would increase the probability of abuse and overdose of opioids by making the wrong decision by the health care providers (Simon et al., 2009).Virtually every stakeholder in the health care industry, including but not limited to patients, physicians, and entire healthcare organizations, rely on the information exchange process to properly perform their functions. Impediments in a HIS's ability to facilitate information exchange can lead to operational disruptions which, within the context of the processes often undertaken by healthcare organizations and professionals, may mean the difference between the life and death of some patients (Grove, Danehy, Consolazio, Lynch, & Mostashari, 2014; Hildebran, et al., 2014; Hillblom, Schueth, Robertson, Topor, & Low, 2014; Griggs, Weiner, & Feldman, 2015). Therefore, interoperability of HIS systems is essential to improve the effectiveness of health care services rendered (Menachemi & Singh, 2012).

In this era of Industrial Revolution 4.0 (IR 4.0), Health Information Exchange (HIE) plays a critical role in enhancing the healthcare industry. HIE enables an improvement of healthcare services as the healthcare organization could access the same information about a patient. Through HIE, the health care providers are able to see the process of data transfer and access all reports from the laboratory as well as radiology that enable the delivery of a better quality of care to patients (Simon et al., 2009). HIE also play a critical role to prevent the incidence of manipulating health's history by the patient. Recent studies have highlighted that currently various prescription drug monitoring systems have been proven effective in the prevention of opioid abuse and misuse, (Gugelmann, Perrone, & Nelson, 2012; Haegerich, 2014; Hagemeier, Murawski, Lopez, Alamian, & Pack, 2014). The success of this monitoring system is commonly due to the interoperability characteristic of the system such as Prescription Drug Monitoring System (PDMP) whereby the physician could fully access the patient real-time information (Hufstader Gabriel et al., 2014). The accessibility of patient real-time information could help the physician to truly monitor the patient's history and hinder the probability of drug abuse among the patients.

The significance of information exchange in healthcare systems goes beyond patient care and doctor's performance; rather it extends to healthcare organization management. It is especially important in-patient care, particularly in monitoring a patient's medical history as well as medication to avoid medical errors associated with lack of information. Information sharing among healthcare organizations and providers is facilitated by interoperability. It ensures that health information regarding patients' medical history, as well as medication, is available and accessible through a central prescription drug monitoring system. Consequently, an organizational health information system improves the effectiveness and efficiency of controlling and managing cases of prescription drug abuse and misuse in healthcare settings.

Currently in Saudi Arabia, although most of the hospital has developed their own HIS, the absence of HIE (interoperability) still could be observed. Although opioid abuse is among the prevalent issues in Saudi Arabia, the current HIS systems do not allow the sharing of pertinent health information with other external HIS, such as those that describe their use of opioids and other controlled substances nature (Al-Zahrani, Eldali and Al-Shahri, 2014 Alshammary et al., 2014). Therefore, due to the lack of HIE, the role of current HIS in curbing opioid abuse's incident in Saudi Arabia is still unknown. Also, the healthcare provider's perception of the ability of HIS in overcoming opioid's abuser is unknown. Therefore, to fill the gap, the present study intended to explore the current HIS system processes in two medical facilities in Saudi Arabia and to identify physician and pharmacist's perception on the role of HIS in curbing the issue of opioid's abuse.

3 MATERIALS AND METHODS

Methodology

Study Design, Setting, Population & Tool

A Case study research was performed at Emergency Department, King Faisal Specialist Hospital and Research Centre (KFSHRC) and Al-Amal Mental Health Complex (AAMHC) from 1st March to 20thApril2018 whereby an observation on the Health Information System (HIS) workflow process was conducted at the selected setting.The observation was specifically focused on the usage of HIS on the prescription of opioid drugs.

Also, a 5 Likert-scale questionnaire was adopted by past studies and was modified according to the suitability of the study settings identified during the observation's period and was distributed to 37 respondentscovering 17 physicians at Emergency Department and 20 pharmacists. These respondents were chosen using the mixture of purposive, convenience and snowballing sampling technique in ensuring the respondents has knowledge and experience in dealing with opioid abused patients. The 5 Likert-scales was used as currently there is no agreement on the number of Likert-scales that needs to be used in the case study research and there were proven to be no reliability difference between the usage of 5 Likert-scales and 7 Likert-scales (Leung, 2011). The questionnaire was designed in two different formats namely for physician and pharmacist. The different format was not intended to compare the difference between physician and pharmacist but, due to the suitability of the working environment and job scope of each profession.

Aims for this study include the following:

1 To determine the current workflow process of the current Health Information System (HIS).

2 To explore the magnitude of the opioid abuse in Saudi Arabia.

3 To determine the general perception of the role of HIS in Saudi Arabia's healthcare among the physician and pharmacist.

4 To explore the physician and pharmacist's perception on the usage of HIS n combating Opioid Abuse.

These aims are further explained in the following sub-sections.

Aim 1: To determine the current workflow process of the current Health Information System (HIS).

Analysis

To understand the current workflow of HIS, this study has conducted observation in both study settings. During the observation, the role of HIS in curbing and managing opioid abuse was particularly observed. The observation was done through frequent visits to both hospitals and at a different shift to meet different people at their convenient time and also to observe the work on different time of the day to see if any different patterns exist. The observation was conducted by focusing on following concerns;

i. Respondent's working environment.

ii. Respondent's perception towards the current role of HIS

iii. The current usageof the HIS by respondents to identify opioid abused patient

iv. Respondent's perception on their usage of current HIS in controlling opioid abuse issues

v. Factors that minimize the usage of HIS in curbing the opioid abused issues

vi. Requirements that they might miss to enhance the use of HIS in mitigating the opioid abused issues

To further explore concerns number i and ii, this study has further developed Aim 2 for concern i and Aim 3 for concern ii. Subsequently, Aim 4 was further developed to answer concern iii to vi.

Aim 2: To explore the magnitude of the opioid abuse in Saudi Arabia.

Theme

In determining the seriousness of opioid abuse in Saudi Arabia, 4 themes were developed from the distributed questionnaire. The themes were:

i. Awareness of abuse issue in Saudi Arabia

ii. Level of significant of opioid's control issue in Saudi Arabia

iii. Importance of patient education

iv. Visible actions and efforts taken to control opioid abuse

Analysis

The collected data from the questionnaire were analyzed descriptively using mean (SD) and the mean score of each profession (physician and pharmacist) were presented individually. Subsequently, a statistical test using Levene's Test was performed to determine the equality of variances of the means between physician and pharmacist. Once the homogeneity of variance was determined, an Independent Sample T-Test was performed to determine the mean difference between both groups of the profession. All analysis was conducted using SPSS version 24 with p-value <0.005 was considered as statistically significant.

Aim 3: To determine the general perception of the role of HIS in Saudi Arabia's healthcare among the physician and pharmacist.

Themes

In exploring the usage of HIS, the questionnaire was divided into 7 themes namely;

i. Technology value in health care
ii. HIS provides cares in health care for patients
iii. Confidence in HIS
iv. Providing efficient health care
v. Used E-Prescription on prescribing medications
vi. Hospital has an electronic system to track and monitor patient
vii. HIS control and minimize drug abuse cases

Analysis

Firstly, the data were analyzed descriptively by calculating the mean (SD) score of each theme. Following the descriptive analysis, the data were analyzed statistically using One-Way Annova to determine the mean difference between physician and pharmacist on their perception towards the functionality of HIS statistically. A test of homogeneity of variance was also performed using Levene's Test to assess the equality of mean's variance between physician and pharmacist. All analysis was conducted using SPSS version 24 with p-value <0.005 was considered statistically significant.

Aim 4: To explore the physician and pharmacist's perception on the usage of HIS in combating Opioid Abuse.

Themes

In exploring the physician and pharmacist's perception of the role of HIS in addressing the issue of opioid abuse, 8 themes were developed namely;

i. Controlling substance

ii. Reduce death

iii. Discourage in dispensing control substance

iv. Doctor Shopping

v. Prescription

vi. Privacy Control

vii. Acceptance of HIS

viii. Acknowledgement of technology

ix. Implementation of an interoperable centralized prescription drug monitoring system

Analysis

The analysis performed in Aim 4 was similar with Aim 3 whereby firstly the data were analyzed descriptively and subsequently One-Way Annova was performed to statistically determine the mean difference between physician and pharmacist on their perception towards the functionality of HIS in combating opioid abuse. A test of homogeneity of variance was again performed using Levene's Test to assess the equality of mean's variance between physician and pharmacist. All analysis was conducted using SPSS version 24 with p-value <0.005 was considered statistically significant.

3. Results

Distribution of Respondent

From the 37 respondents, 67.6% of them were male and 32.4% were female. The majority (37.8%) of the respondents were aged between 31 years old to 40 years old and more than 50.0% of the respondent were among the non-Saudi. The distribution of the respondent also showcased that more than 50.0% of the respondents were the pharmacist. The distribution of the respondent is summarized in Table 2.

Table ' Distribution of Respondent

Variable	Frequency	%
Gender		
Male	25	67.6
Female	12	32.4
Age Group		
20 - 30	5	13.5
31 - 40	24	37.8
41 - 50	12	32.4
> 50	6	16.2
Nationality		
Saudi Arabian	14	37.8
Non-Saudi	23	62.2
Profession		
Physician	17	45.9
Pharmacist	20	54.1
Year of Experience		
5 - 10	18	48.6
11 - 15	8	21.6
16 - 20	9	24.3
>20	2	5.4

Output for Aim 1: *Observation of Current HIS in Saudi Arabia*

From the study's observation, it was identified that King Faisal specialist hospital is a fully automated workflow involving HIS but within the organization only. On the other hand, Al-Amal Mental Health Complex is partially automated, the usage of paper in certain parts of the processes could be observed. Figure 1 represents the current process with emphasis on the process of identifying the drug abuse patients from physician and pharmacist perspective. From Figure 1, it is apparent that HIS is being fully used to register a patient, access the medical record of the patient, access medication orders catalogue and to order opioid or other controlled drugs.

Although there was a slight difference in the work process of HIS in both hospitals, the observation has identified 5 common critical issues in both facilities. Firstly, the current HIS in both hospitals could not be integrated with the other health care facilities for information sharing purposes. Secondly, the study's observation has concluded that the current HIS system was not user friendly and has caused redundancy of opioid prescriptions. For instance, there is a policy restriction to access the system in an impromptu manner for abuse suspicion. Thirdly there was also a restriction on the interoperability of the system at both sides.

Currently, the physician could only utilize the information from the current HIS solely within their facilities which the patient might have a history of abusing the drug in other healthcare facilities but could not be traced with the current HIS. Also, limited communication between physician and pharmacist was observed when prescribing opioids repeatedly to patients which preventing the current HIS's role to curb the issue of abused opioids. From the observation, it also could be concluded that as the true capability of HIS in curbing opioid drugs abuse is still unknown, interest to develop an interoperable information system is still low relating to low funding to develop an interoperable information system.

Figure 1 HIS Workflow

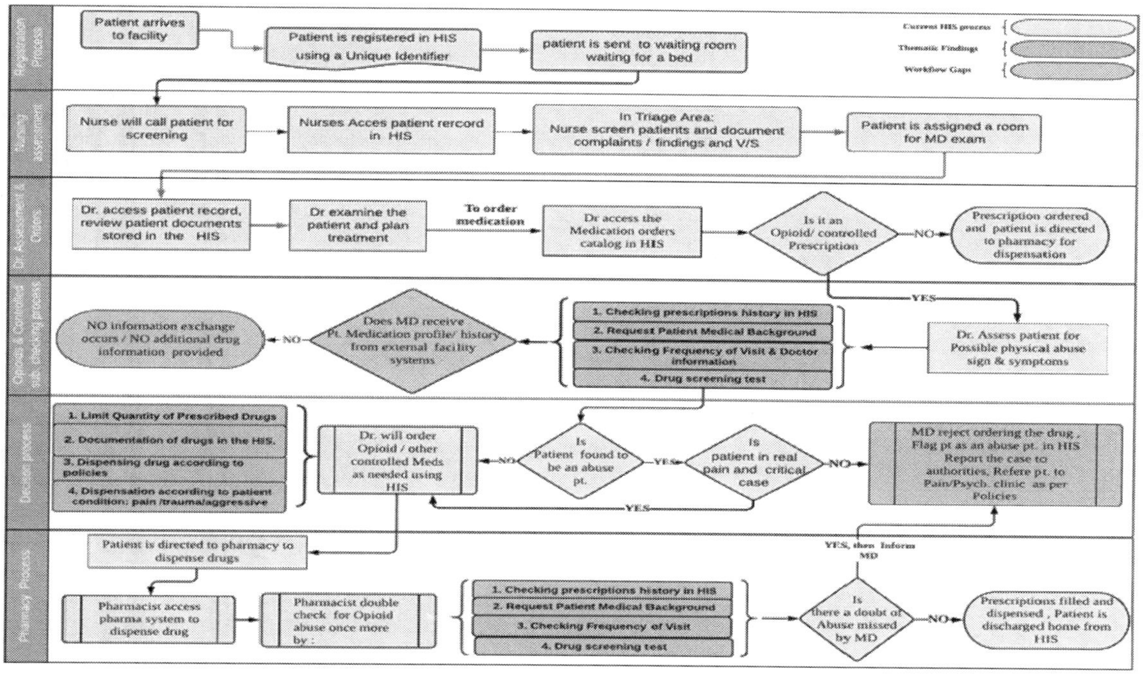

Output for Aim 2:*Magnitude of Opioid Abuse in Saudi Arabia*
A five-point Likert scale type of questions was used in the questionnaire which implies the lowest mean score that could be obtained per theme is 1 and the highest is 5. The output of the mean scores across all four themes was shown to be more positive than negative whereby the value of mean for each theme equals more than 3 (Table ٣). This further indicates that although there was a slight variation in the mean scores between each profession, both physician and pharmacist has observed the opioid abuse as a prevalent issue in Saudi Arabia. While there were definitely differences between the responses of the two groups, the difference was proven to be statistically insignificant and those differences were marginal at best. More importantly, the findings from the Levene's Test for Equality of Variances, show that the physicians and pharmacists carried a similar magnitude of opioid abuse in Saudi Arabia whereby no statistically significant variance was identified on the mean score.This implies that both professions have a similar level of awareness and perception of opioid abuse and misuse issues in the country. The output of the statistical analysis is presented in Table ٣.
Table ٣*Mean* Score for Magnitude of Opioid Abuse

Theme	Profession	N	Mean	Std. Deviation	Std. Error Mean
Aware of Abuse issue in Saudi Arabia	Physician	17	4.18	0.951	0.231
	Pharmacist	20	4.35	0.745	0.167
Control of substance is a significant issue	Physician	17	4.47	0.717	0.174
	Pharmacist	20	4.40	0.754	0.169
Patient Education	Physician	17	3.94	1.088	0.264
	Pharmacist	20	3.90	1.165	0.261
Visible actions and efforts	Physician	17	3.53	0.943	0.229
	Pharmacist	20	3.90	0.912	0.204

Table 7 Statistical Analysis for Aim 2

Theme		Levene's Test for Equality of Variances		t-test for Equality of Means						
		F	Sig.	t	Df	Sig. (2-tailed)	Mean Difference	Std. Error Difference	95% Confidence Interval of the Difference	
									Lower	Upper
Aware of Abuse issue in Saudi Arabia	Equal variances assumed	0.792	0.380	-0.622	35	0.538	-0.174	0.279	-0.74	0.393
	Equal variances not assumed			-0.610	30.144	0.547	-0.174	0.285	-0.755	0.407
Control of substance is a significant issue	Equal variances assumed	0.032	0.859	0.290	35	0.773	0.071	0.243	-0.423	0.564
	Equal variances not assumed			0.291	34.521	0.773	0.071	0.242	-0.422	0.563
Patient Education	Equal variances assumed	0.380	0.542	0.110	35	0.913	0.041	0.373	-0.716	0.798
	Equal variances not assumed			0.111	34.662	0.912	0.041	0.371	-0.712	0.794
Visible actions and efforts	Equal variances assumed	0.162	0.690	-1.213	35	0.233	-0.371	0.306	-0.991	0.25
	Equal variances not assumed			-1.209	33.641	0.235	-0.371	0.306	-0.994	0.252

Output for Aim 3: *General Perception on the Functionality of HIS*

The output from the mean score of the questionnaire indicates that both of the healthcare professionals have a positive perception of the general functionality of HIS (Table 5). For instance, both physician and pharmacist had a good perception on the role of HIS in providing efficient healthcare with the mean score of 4.94 (SD: 0.243) and 4.85 (SD:0.366), respectively. However, in line with the output during the observation at study setting, most of the respondents perceive that the current HIS is still lacking in monitoring and tracking patient's drugs history. The overall mean score was 2.73 (SD: 1.895) out of 5 and only 2.71(SD:1.929) for the physician and 2.75 (SD: 1.916) for the Pharmacists.

Overall, although there were slight differences in mean score between both professions, the output from the ANOVA showed that the differences were statistically insignificant. This indicates the general perception of the functionality of HIS was aligned between both professions. This is further proven by the output of the test of Homogeneity of Variance whereby mostly the mean score between both professions showed insignificant variance. However, the test of Homogeneity of Variance showed that there is a variation in the respondents' opinion about the functionality of E-prescription tool within the HIS in use.

Table 4 Statistical Analysis for Aim 3

Theme	Profession	Annova*				Levene Test	
		N	Mean	Std. Deviation	Std. Error	F	p value
Technological Value in Health Care	Physician	17	4.35	0.996	0.242	1.797	0.189
	Pharmacist	20	4.20	0.768	0.172		
	Total	37	4.27	0.871	0.143		
HIS provides cares in Health care for patients	Physician	17	4.53	0.717	0.174	0.031	0.862
	Pharmacist	20	4.55	0.759	0.170		
	Total	37	4.54	0.730	0.120		
Confidence in HIS	Physician	17	4.35	0.862	0.209	0.423	0.519
	Pharmacist	20	4.40	0.681	0.152		
	Total	37	4.38	0.758	0.125		
Providing Efficient Healthcare	Physician	17	4.94	0.243	0.059	3.377	0.075
	Pharmacist	20	4.85	0.366	0.082		
	Total	37	4.89	0.315	0.052		
E-Prescription Systems	Physician	17	4.29	1.312	0.318	5.589	0.024
	Pharmacist	20	4.70	0.470	0.105		
	Total	37	4.51	0.961	0.158		
Hospital has electronic system to track, monitor patient	Physician	17	2.71	1.929	0.468	0.016	0.901
	Pharmacist	20	2.75	1.916	0.428		
	Total	37	2.73	1.895	0.312		
HIS Control and minimize abuses	Physician	17	4.65	0.606	0.147	1.194	0.282
	Pharmacist	20	4.55	0.759	0.170		
	Total	37	4.59	0.686	0.113		

*p value for Annova Test were all >0.05

Output for Aim 4 : *Perception on the role of HIS in Combating Opioid Abuse*
As evidence by the mean scores, a positive perception of the role of HIS in combating opioid abuse within both professions could be observed. The mean scores of each theme indicate a score of more than 3. However, the only exception was for the theme of Innovation of Privacy whereby the mean score was low than 3.
The statistical significance of these findings was later on confirmed using the Levene's Test of Homogeneity of Variances and One-Way ANOVA. The analysissuggested that the actual integration of technology in the current Saudi Arabian HIS differs from the way how the doctors and physicians perceived it to have an influence over the prescription drug abuse and misuse monitoring and treatment outcomes in Saudi Arabia. It is worth noting, however, that the participants' perceptions about all the other aspects of interoperability and how it can help mitigate opioid abuse and misuse in Saudi Arabia were generally favourable but issues such as the privacy of the patient require more attention as both professions perceive this theme negatively with an overall mean score of 2.14 (SD: 1.294).

Table ᵃ Statistical Analysis for Aim 4

Theme	Profession	Annova*				Levene Test	
		N	Mean	Std. Deviation	Std. Error	F	P value
Integration Technology helps in control substance in Saudi Arabia	Physician	17	4.47	0.624	0.151	1.293	0.174
	Pharmacist	20	4.40	0.503	0.112		
	Total	37	4.43	0.555	0.091		
Implementing an Interoperable, Centralized Prescription drug to reduce death	Physician	17	4.29	0.920	0.223	3.066	0.089
	Pharmacist	20	4.60	0.598	0.134		
	Total	37	4.46	0.767	0.126		
Discourage to dispense control substance Discourage	Physician	17	4.53	0.874	0.212	2.480	0.124
	Pharmacist	20	4.20	1.361	0.304		
	Total	37	4.35	1.160	0.191		
Doctor shopping	Physician	17	4.65	0.702	0.170	0.135	0.716
	Pharmacist	20	4.70	0.733	0.164		
	Total	37	4.68	0.709	0.117		
Prescription Implementation of drug	Physician	17	4.41	0.870	0.211	4.121	0.050
	Pharmacist	20	4.65	0.489	0.109		
	Total	37	4.54	0.691	0.114		
Innovation of Privacy if implemented	Physician	17	2.06	0.899	0.218	9.850	0.003
	Pharmacist	20	2.20	1.576	0.352		
	Total	37	2.14	1.294	0.213		
Positive Feedback using HIS	Physician	17	4.47	0.514	0.125	0.054	0.818
	Pharmacist	20	4.55	0.510	0.114		
	Total	37	4.51	0.507	0.083		
Technology is a must	Physician	17	4.53	0.514	0.125	8.761	0.005
	Pharmacist	20	4.80	0.410	0.092		
	Total	37	4.68	0.475	0.078		
Interoperable centralize prescription drug monitoring system implemented	Physician	17	4.65	0.702	0.170	17.204	0.000
	Pharmacist	20	4.95	0.224	0.050		
	Total	37	4.81	0.518	0.085		

*p value for Annova
Test were all >0.05

Discussions

Overall, there was a clear evidence thatthe current interoperability of the current HIS was sufficient within the organization. However, the HIE does not go beyond the organization whereby the level of integration between one or more sites was insufficient for both the KFSHRC and AAMHC. This has prevented the healthcare practitioners to do their respective jobs, as far as identifying, managing, treating, and preventing patients with opioid abuse and misuse were concerned. This finding echoed past studies that highlighted the absence of interoperability in facilities, HIS, and their various functions which prevent the healthcare facilities to provide the best quality of healthcare treatment to the patient(Bonobaid, Fan, & Almeziny, 2016; Assuli, Shabtai, & Leshno, 2013; Hobson, 2017; Kasthurirathne, Mamlin, Kumara, Grieve, & Biondich, 2015).

Finding on the physician and pharmacist's perception on the magnitude of opioid's abuse and misuse in Saudi Arabia showed that the misuse of opioid is serious problem and caught the concerned of the physicians and pharmacists. As reported by past studies, opioid drugs are being widely used in Saudi Arabia specifically in treating the pain of chronic disease's patients (Abdulla et al., 2013; Manglik et al., 2016). The high usage of opioid drugs without a close and proper monitoring of its usage among the patients may relate to the incidence of opioid's abuse among the patients. The similar trend has been observed in past study whereby an increasing trend of opioid's abuse incidence could be identified when the sale of opioid's drug was increased % (Jones, Mack and Paulozzi, 2006; Worley, 2012).

From the questionnaire on physician and pharmacist perception of the current HIS, it's apparent that, both professionals perceived HIS positively in overcoming the issue of drug abuse. This finding was consistent with previous finding by Hanh (2011) that explores the strategies that could mitigate the risk of opioid abuse. Hanh (2011) has highlighted that prescription monitoring program through HIS may be beneficial in reducing the incidence of drug abuse especially opioid. However, without having an interoperable HIS, the collected data may be meaningless to closely monitor the patient. Evidently, in this study, the mean score of both professionals showed a negative perception on the ability of the current HIS in monitoring and tracking patients drug history. This negative perception could be associated with the flaw of the current HIS namely the absence of interoperability that has become a huge barrier for health information exchange.

The lack of health information exchange within healthcare facilities may hinder the healthcare practitioner's in curbing the opioid's abused issues. For instance, past study has highlighted that the incidence of doctor shopping (abuse patients visiting multiple health care facilities seeking the same drug) is highly contributed by the lack of real-time information exchanged within healthcare facilities (Worley, 2012; Patrick *et al.*, 2016; Buchmueller and Carey, 2018). The absence of health information exchange may relate the patient to manipulate their drug prescription's history and schedule as the accessibility of patient's medical history by a physician is limited (Lehne *et al.*, 2019). Therefore, integrated HIS plays a crucial role in curbing the opioid abused issue.

Echoing the past study, in this study, both pharma

cist and physician carried a positive perception towards the role of HIS in curbing abuse of opioids, but this remain within theirorganization only. However, looking at the observation on the workflow of the current HIS, the lack of interoperability of HIS between study settings and other facilities could become the major obstacle for the current HIS to plays its role in mitigating opioid's abuse issues. The same finding was identified in a study by Hagemeier et al.(2014) thatexamined the pharmacists' perceptions regarding opioid pain reliever abuse and the role that interoperability plays in mitigating such issues. The absence of interoperability may hinder the healthcare practitioner to closely monitor the drug prescription history of the patient that may cause difficulties for them to curb the issues of abuse and misuse of opioid. The negative perception on the ability of the current HIS to track and monitor patient could be observed under Aim 3 whereby the mean score of physician and pharmacist on the ability of the current HIS to track and monitor was recorded to be less than 3. This has further confirmed the importance of interoperability to monitor and track patient's history. Only through real-time health information exchange the health care providers could mitigate the incidence of opioid abuse. The importance of interoperability in addressing opioid medication abuse and misuse issues were also clearly established in other studies, backed by their research results and findings(JM, 2016; BD, 2015; Bert, Galis, Passi, Gualano, & Siliquini, 2015; Bloomfield & Mandel, 2017; Anderson, Zlateva, Khatri, & Ciaburri, 2015).

At Aim 4, physicians and pharmacists concerned on the privacy of the patient information through using HIS to overcome opioid abused could also been observed. The physician and pharmacist's concern on privacy issue echoed the concern by past studies as this issue has been frequently debated by past studies. For instance, a study by Brown et al. (2017) has highlighted that in implementing drug monitoring program through HIS, patient's information might need to be compromised to reach the optimal result of the drug monitoring program (Brown *et al.*, 2017). Also, another study by Marco et al. (2016) has highlighted that one of the challenges to monitor patient's drug history using HIS is the confidentiality issue which is deemed as an unresolved ethical issue.

Study Limitations

The researcher has identified four key limitations of the study that was conduted. The first one is that the study focused only on the Opioid abuse and misuse issue. Nevertheless, the findings of this study arebelieved to be relevant to be referred in combating other controlled substances too. The second limitation was the required use of classified and confidential medical information during the study and the stringent security and confidentiality rules in the health care industryas well as other data centers. The third limitation was the lack of similar or relatively similar articles and studies about opioid abuse and misuse specific to Saudi Arabia. Lastly, due to the time constrain, this study only capable to explore the opioid abuse issue within two facilities only.

Conclusions

As the issue of opioid abuse is a prevalent issue within Saudi Arabia, it's highly important to address and mitigate this issue. The issue of opioid abuses could be resolved through close tracking and monitoring of the patient's medical history using the HIS. However, although both study settings have developed their own HIS, there is still an absence of interoperability which hinder the physicians and pharmacists effort to curb the issue of opioid abuses using the current HIS. The current workflow processes being used in some of its biggest health care institutions are already outdated, as evidenced by the low levels of interoperability in many of their functions. In the era of IR 4.0, health information exchange plays a crucial role to determine the effectiveness of healthcare treatment rendered to the patient. The lack of health information exchange may cause misjudgment during the treatment session. This absence of interoperability may cause a higher incidence of opioid abuse or misuse in Saudi Arabia as the real-time information of patient's medication history could not be gathered using the current HIS.

Recommendations

As to date, research on the impact of HIS towards the issue of opioid abuse is still limited specifically in Saudi Arabia. Therefore more research on this kind of subject is highly needed and recommended especially involving the ownership and responsibility of the process support system that crosses the organization's boundaries. This kind of research is believed could provide support towards the patient centered healthcare. In addition, research on the improvement of the function of the current HIS to curb issues such as opioid abuse also needs to be increased to make the patient electronic medical record easily available across any organization without compromising and violating the patient's privacy.

References

Albooshi, H., Hulse, G., Kashef, A., Al Hashmi, H., Shawky, M., Al Ghaferi, H., . . . Tay, G. (2016). The patterns of substance use disorders in the United Arab Emirates in 2015: Results of a national rehabilitation centre cohort study. *Substance Abuse, Treatment, and Prevention Policy*, 11-19.

Alshammary, S. *et al.* (2014) 'Palliative care in Saudi Arabia: Two decades of progress and going strong', *Journal of Health Specialties*, 2(2), p. 59. doi: 10.4103/1658-600x.131749.

Al-Zahrani, O., Eldali, A. and Al-Shahri, M. Z. (2014) 'Prevalence and severity of pain in cancer patients in an outpatient palliative care setting in Saudi Arabia', *Qatar Medical Journal*, 2014(1), pp. 38–45. doi: 10.5339/qmj.2014.6.

Anderson, D., Zlateva, I., Khatri, K., & Ciaburri, N. (2015). Using health information technology to improve adherence to opioid prescribing guidelines in primary care. *The Clinical Journal of Pain.*

Assuli, O., Shabtai, I., & Leshno, M. (2013). The impact of EHR and HIE on reducing avoidable admissions: controlling main differential diagnoses. *BMC Medical Informatics and Decision Making.*

BD. (2015). Optimizing medication safety: At King Abdullah Medical City advanced technologies help improve safety, security and control of N&C medications. *Evsins, Switzerland: Becton, Dickinson, and Company.*

Bert, F., Galis, V., Passi, S., Gualano, M., & Siliquini, R. (2015). Differences existing between USA and Europe in opioids purchase on Internet: An Interpretative Review. *Journal of Substance Abuse*, 200-207.

Bloomfield, W., & Mandel, M. (2017). Opening the Duke electronic health record to apps: Implementing SMART on FHIR. *International Journal of Medical Informatics*, 1-10.

Bonobaid, S., Fan, I., & Almeziny, M. (2016). Investigation Interoperability Problems in Pharmacy Automation: A Case Study in Saudi Arabia. *Procedia Computer Science*, 329-338.

Brown, R. *et al.* (2017) 'Impact of New York prescription drug monitoring program, I-STOP, on statewide overdose morbidity', *Drug and Alcohol Dependence.* doi: 10.1016/j.drugalcdep.2017.05.023.

Buchmueller, T. C. and Carey, C. (2018) 'The effect of prescription drug monitoring programs on opioid utilization in medicare', *American Economic Journal: Economic Policy.* doi: 10.1257/pol.20160094.

Centers for Disease Control (2011) *Vital signs: overdose of prescription opioid pain relievers-United State, 1999 - 2008, MMWR.*

Denisco, R., Kenna, G., O'Neil, M., Kulich, J., Moore, P., Hersh, E., & Katz, N. (2011). Prevention of prescription opioid abuse. *The Journal of the American Dental Association*, 800-810.
Denzin, N., & Lincoln, Y. (2000). Handbook of Qualitative Research. *Newberry Park: Sage*.
Gandhi, R. *et al.* (2016) 'Into the Data Lake', pp. 1–7. Available at: https://www.bcg.com/de-de/publications/2016/big-data-advanced-analytics-technology-digital-look-before-you-leap-into-data-lake.aspx.
Griggs, C., Weiner, S., & Feldman, J. (2015). Prescription drug monitoring programs: examining limitations and future approaches. *Western Journal of Emergency Medicine*.
Grove, D., Danehy, L., Consolazio, M., Lynch, K., & Mostashari, F. (2014). A national study of challenges to electronic health record adoption and meaningful use. *Medical Care*, 144-148.
Gugelmann, H., Perrone, J., & Nelson, L. (2012). Windmills and Pill Mills: Can Prescription Drug Monitoring Programs Tilt the Prescription Drug Epidemic. *Journal of Medical Toxicology*, 378-386.
Hahn, K. L. (2011) 'The Scope of the Problem Abuse of Prescription Opioid Drugs in the United States', *\Am Health Drug Benefits.*, 2(April), pp. 107–114. Available at: www.AHDBonline.com.
Haegerich, T. (2014). What We Know, and Don't Know, about the Impact of State Policy and Systems level Interventions on Prescription Drug Overdose. *Drug Alcohol Depend.*
Hagemeier, N., Murawski, M., Lopez, N., Alamian, A., & Pack, R. (2014). Theoretical exploration of Tennessee community pharmacists' perceptions regarding opioid pain reliever abuse communication. *Research in Social and Administrative Pharmacy*, 562-575.
Halsted, M., & Froehle, M. (2008). Design, Implementation, and Assessment of a Radiology Workflow Management System. *American Journal of Roentgenology*, 321-327.
Hildebran, C., Cohen, D., Irvine, J., Foley, C., O'Kane, N., Beran, T., & Devo, R. (2014). How Clinicians Use Prescription Drug Monitoring Programs: A Qualitative Inquiry. *Pain Medicine*, 1179-1186.
Hillblom, D., Schueth, A., Robertson, S., Topor, L., & Low, G. (2014). The impact of information technology on managed care pharmacy: today and tomorrow. *Journal of Managed Care Pharmacy*, 1073-1079.
Hobson, C. (2017). Achieving your FHIR share of interoperability. *Health Management Technology*.
Hufstader Gabriel, M. *et al.* (2014) 'Adoption of electronic prescribing for controlled substances among providers and pharmacies', *The American journal of managed care*.
JM. (2016). The Effects of Low Self-Control and Delinquent Peers on Alcohol, Tobacco, and Drug Use in a Sample of Saudi Arabian Youth. *Int J Offender Ther Comp Criminol*, 1569-1587.
Jones, C. M., Mack, K. A. and Paulozzi, L. J. (2006) 'Pharmaceutical Overdose Deaths, United States, 2010', *JAMA - Journal of the American Medical Association*, pp. 524–525. doi: 10.3122/jabfm.19.5.524.

Kasthurirathne, S., Mamlin, B., Kumara, H., Grieve, G., & Biondich, P. (2015). Enabling better interoperability for healthcare: lessons in developing a standards based application programing interface for electronic medical record systems. *Journal of Medical Systems*.
Keller, M., Kelling, S., Cornelius, D., Oni, H., & Bright, D. (2015). Enhancing practice efficiency and patient care by sharing electronic health records. *Perspectives in Health Information Management*.
Kellermann, A., & Jones, S. (2013). What it will take to achieve the as-yet-unfulfilled promises of health information technology. *Health Affairs*, 63-68.
Khan, W., Hussain, M., Latif, K., Afzal, M., Ahmad, F., & Lee, S. (2013). Process Interoperability in Healthcare Systems with Dynamic Semantic Web Services. *Computing*, 1-26.
Kohli, R., & Tan, S. (2016). Electronic Heath Records: How Can IS Researchers Contribute to Transforming Healthcare. *MIS Quarterly*, 553-574.
Lehne, M. *et al.* (2019) 'Why digital medicine depends on interoperability', *npj Digital Medicine*. Springer US, 2(1), pp. 1–5. doi: 10.1038/s41746-019-0158-1.
Lupton, D. (2014). Critical perspectives on digital health technologies. *Sociology Compass*, 1344-1359.
Menachemi, N., & Singh, S. (2012). Health Information Technology in the International Context. *Bingley, UK: Emerald Group Publishing*.

Miller, K., & MacCaull, W. (2009). Toward web-based careflow management systems. *Journal of Emerging Technologies in Web Intelligence Special Issue: E-Health Interoperability*, 137-145.
Namli, T., & Dogac, A. (2010). Testing Conformance and Interoperability of eHealth Applications. *Methods of Information in Medicine*, 281.
Patrick, S. W. *et al.* (2016) 'Implementation of prescription drug monitoring programs associated with reductions in opioid-related death rates', *Health Affairs*. doi: 10.1377/hlthaff.2015.1496.
Simon, S. R. *et al.* (2009) 'Patients' attitudes toward electronic health information exchange: qualitative study.', *Journal of medical Internet research*, 11(3), pp. 1–8. doi: 10.2196/jmir.1164.
Worley, J. (2012) 'Prescription drug monitoring programs, a response to doctor shopping: Purpose, effectiveness, and directions for future research', *Issues in Mental Health Nursing*. doi: 10.3109/01612840.2011.654046.
Yin, R. (2011). Qualitative Research from Start to Finish. *New York, NY: Guilford Press.*
Zhang, J., Lu, H., Nie, Z., Hung, Z., & Aalst, W. (2009). Radiology Information System: A Workflow Based Approach. *International Journal of Computer Assisted Radiology and Surgery*, 509-516.

Made in United States
Orlando, FL
04 November 2026

90201013R00020